THE POETRY OF BISMUTH

The Poetry of Bismuth

Walter the Educator

Silent King Books

SILENT KING BOOKS

SKB

Copyright © 2024 by Walter the Educator

All rights reserved. No part of this book may be reproduced in any manner whatsoever without written permission except in the case of brief quotations embodied in critical articles and reviews.

First Printing, 2024

Disclaimer
This book is a literary work; poems are not about specific persons, locations, situations, and/or circumstances unless mentioned in a historical context. This book is for entertainment and informational purposes only. The author and publisher offer this information without warranties expressed or implied. No matter the grounds, neither the author nor the publisher will be accountable for any losses, injuries, or other damages caused by the reader's use of this book. The use of this book acknowledges an understanding and acceptance of this disclaimer.

"Earning a degree in chemistry changed my life!"
- Walter the Educator

dedicated to all the chemistry lovers, like myself, across the world

BISMUTH

Atoms dance and elements prance,

BISMUTH

There lies a metal, bismuth, in a mystical trance.

BISMUTH

With colors like a rainbow, it catches the eye,

BISMUTH

A captivating element, soaring high in the sky.

BISMUTH

Upon the periodic table, it finds its place,

BISMUTH

Amongst the heavy metals, with grace and embrace.

BISMUTH

But bismuth is unique, it's not like the rest,

BISMUTH

With crystals forming patterns, it's truly the best.

BISMUTH

Its atomic number, eighty-three in line,

BISMUTH

With electrons swirling, a spectacle so fine.

BISMUTH

Its nucleus holds protons, neutrons in tow,

BISMUTH

A symphony of particles, putting on a show.

BISMUTH

In nature's laboratory, bismuth does dwell,

BISMUTH

In ores and minerals, where its secrets swell.

BISMUTH

Extracted with care, from the earth's deep embrace,

BISMUTH

To be shaped and molded, with finesse and grace.

BISMUTH

Oh bismuth, how you shimmer, with iridescent gleam,

BISMUTH

In shades of pink, blue, and yellow, like a vivid dream.

BISMUTH

Your crystal lattice structure, a marvel to behold,

BISMUTH

A testament to nature's artistry, precious like gold.

BISMUTH

In laboratories, scientists study your ways,

BISMUTH

Exploring your properties, in intricate arrays.

BISMUTH

From alloys to medicines, you play a vital role,

BISMUTH

Innovating industries, with each new goal.

BISMUTH

But beyond the confines of science and lore,

BISMUTH

Bismuth holds a mystique, an allure to explore.

BISMUTH

For in its essence lies a deeper truth,

BISMUTH

A connection to the cosmos, eternal in youth.

BISMUTH

From the stars we came, and to the stars we'll return,

BISMUTH

Bismuth whispers softly, a lesson to learn.

BISMUTH

In its quiet presence, we find solace and peace,

BISMUTH

A reminder of our journey, where all struggles cease.

BISMUTH

So let us cherish bismuth, in all its splendid form,

BISMUTH

A testament to nature's beauty, amidst the storm.

BISMUTH

For in its humble existence, we find our place,

BISMUTH

A cosmic dance of atoms, bound by grace.

BISMUTH

With each crystalline structure, a tale untold,

BISMUTH

Of atoms aligning, in patterns manifold.

BISMUTH

Bismuth's beauty transcends the earthly realm,

BISMUTH

A cosmic masterpiece, at the helm.

BISMUTH

ABOUT THE CREATOR

Walter the Educator is one of the pseudonyms for Walter Anderson. Formally educated in Chemistry, Business, and Education, he is an educator, an author, a diverse entrepreneur, and he is the son of a disabled war veteran. "Walter the Educator" shares his time between educating and creating. He holds interests and owns several creative projects that entertain, enlighten, enhance, and educate, hoping to inspire and motivate you.

Follow, find new works, and stay up to date with Walter the Educator™ at WaltertheEducator.com

www.ingramcontent.com/pod-product-compliance
Lightning Source LLC
LaVergne TN
LVHW010412070526
838199LV00064B/5271